The Gray Field Uniform of the German Army

The Gray Field Uniform of the German Army

With numerous lithographic color illustrations

Moritz Ruhl

Illustrations by F. Hubner

English translation by Patrick C. O'Driscoll

ISBN: 978-1-7357542-0-8

TRANSLATOR'S DEDICATION

To my wife Elaine

CONTENTS

Forward

Any translator must make decisions on how best to express what the original author was trying to communicate. No translation can be the same word for word as the original.

I have tried to stay as close to the original sentences as possible after adjusting for grammar and German words that have no direct translation in English, particularly when it comes to words for specific uniform components.

I have used the English versions of terms that are virtually identical to the German term albeit with a German spelling and pronunciation. For example, Cuirassier, Hussar, Foot Guards etc. are used rather than their German counterparts. The English translation is merely an anglicized version of the term and using it for the English reader enhances clarity.

In the case of German terms, for which the best English translation could still leave ambiguity as to what is being discussed such as, jäger, schützen, waffenrock etc. or German words in the English lexicon like Reich, the German word was used.

Explanations of the meaning of these terms are given in the footnotes and in the glossary of German words in the appendix.

The Gray Field Uniform of the German Army

Through the all highest cabinet order of 23 February 1910 and the decree of 18 March 1910, of changes decided for the German army, in addition to the present uniforms, a new field gray uniform (for jäger and schützen a gray green uniform) is established.

Occasions for which the dark blue uniform or the field gray uniform are worn are given in the simultaneously published dress regulations, authorized exactly as follows:

The **dark blue**[1] **Uniform** must be worn:

a) as Parade Dress,
b) in Garrison Guard Duty,
c) for Church Dress,
d) for Court Duty,
e) as Walking Out and Evening Dress.

The **field gray Uniform** must be worn:
a. in the field
b. for all combat- etc. exercises against one another as sham opponents.

For all other occasions it remains for the commanding officer to decide, whether the dark blue or the field gray uniform is used.

---

[1] The description "dark blue Uniform" is understood to be all uniforms of the former color.

Description of the Officers Field Uniform

The regulations given heretofore, are how all, made to order, officers uniforms, are to be manufactured. They must be completely in line with, and the same as, particularly in color, the other ranks uniforms. Above all they are not permitted to be lighter than them.

The Waffenrock (Feldrock)

In General

The **waffenrock** is of the same cut as those of the other ranks, it must fit comfortably since undergarments are intended to be able to be worn beneath it. The waffenrock skirt must extend only halfway down the seat for mounted troops, however for dismounted troops the seat should be entirely covered, so that for the latter it extends 10.5 cm below the crotch.

The waffenrock has one row of eight matt crown buttons; they are all either gold or silver.

At the front on the lower part of the skirt are two slanted pockets of curved cut and descending toward the back. The flaps are rounded without buttons. On the back of the waffenrock are scalloped vertical pocket flaps with six buttons and colored piping, the same as it is on the dark blue uniform; to both sides of the upper two buttons, in alignment with them at waist height, are two hooks.

On the inside of the waffenrock, likewise at waist height, a draw string is attached.

The fabric is broadcloth or doeskin, it is however also permitted now for the field waffenrock to be of tricot (cavalry twill). For jägers and schützen as well as for the machinegun units and the jäger zu pferd (mounted jäger) the same applies however, the fabric is of gray green color. For all other branches it is of field gray color.

The lining of the waffenrock is the fabric color, and the sleeve lining is a lighter color.

The cuffs of all waffenrocks have a slit, no matter what kind of cuff they are. The Guards have a red piping on the edge of the slit.

The **cuffs** as well as the cuff flaps are from the fabric of the waffenrock, they are of the same pattern (therefore Brandenburg or Swedish etc.) as they are on the dark blue uniform. The piping on the cuffs and the cuff flaps respectively are the same color as on the cuffs of the dark blue waffenrock.

The **collar** of the waffenrock, is closed with only two hooks and eyes, and is required to be low enough to make possible the free movement of the head and not hinder it.

For the waffenrocks of the cuirassiers, dragoons and jägers zu pferde the **stand collar** is prescribed, for all other branches there is a **stand and fall collar** with somewhat rounded corners. It shall be worked soft so that it can be a turned up; the upper and lower collar is the same fabric as the waffenrock. Around the collar runs piping of the same color and material as the dark blue uniform.

Collar patches on the field gray uniform are worn in the following manner: When the dark blue officers uniform has embroidery, there are colored collar patches on the field gray waffenrock of the officer (75mm long, 43mm high). In all other instances the waffenrock collar has no patches.

The collar patches correspond to the regulations for the officers litewka[2] regarding piping, materials (cloth or velvet) and the color. However, there are no buttons on the patches. The only exception to these rules is for the Regiments Guard du Corps and the Guard Cuirassiers.

---

[2] Officer's double-breasted undress jacket.

The **Color of the Collar Patches** are:

Generals: poppy red.

General and Flank Adjutants, Generals a' la Suite of his Majesty the Kaiser: silver with poppy red piping
Personal Adjutants of Princes: poppy red.

War Ministry and General Staff: carmine red.

Infantry and all those not given a particular department, institute etc.: from the color of the shoulder straps.

Jägers: dark green

Guard Jäger and Guard Machinegun Unit Nr. 1: dark green with poppy red piping.

Guard Schützen and Guard Machinegun Unit Nr. 2: of green velvet with poppy red piping.

Guard du Corps: poppy red

Guard Cuirassiers: cornflower blue

Dragoons: cornflower blue

Hussars: from the cloth color of the attila with piping as it is on the cap crown.

Uhlans (Lancers): from the color of the epaulette field.

Field Artillery: from the color of the shoulder straps with piping of black velvet.

Foot Artillery: of black velvet with lemon yellow piping.

Engineer and Pioneer Corps, as well as Transportation Troops: of black velvet with poppy red piping.

Supply Train[3]: light blue.

Technical Institute: light blue.

Riding Field Jäger Corps: dark green.
Land and Field Gendarmerie: cornflower blue with dark green piping.

Guard Militia Cavalry: from the color of the epaulette field with dark blue piping.

Medical Officers: dark blue with poppy red piping.

Veterinary Officers: black with carmine red piping.

Litzen are worn on the collar patches of the officer's field waffenrock only when litzen are regulation for the other ranks of those units. In those cases, litzen are also worn on the cuffs.

When a corresponding other ranks uniform does not exist, only the colored collar patches without litzen are worn. This, therefore, is the case for officers of the war ministry, the general staff, the engineer corps, for the mounted field jäger corps, as well as medical and veterinary officers. In contrast the officers of line pioneer battalions and telegraph battalions Nr. 2 and 7, have no collar patches on the field waffenrock.

The litzen are made of silver, matt fabric and are formed exactly as the enlisted men's litzen, in the middle of the main part is found an approximately 3mm wide flash, of matt silver chain embroidery.

The litzen are sewn on the collar patches so that they are, in the case of double litzen, with the fluted ends of the upper and lower litze connected to each other, whereas for the single and old Prussian litzen the edge of the main part is equal width from the upper and lower edge of the patch. For all three litzen types they are 2mm from the patch edge at the front and back. On the cuffs

---

[3] Refers to "Train" in glossary of German terms.

the litzen are sewn on as they are on the other rank's uniforms.

The fastening of the shoulder pieces to the field waffenrock is the same as previously; the epaulette holders, however, are not present.

Adjutant's sash and feldbinden (braided belts) are worn with the field uniform. In combat a gray cover is provided, or in the absence of such, it is not worn entirely. Then in place of the feldbinde, a leather belt is worn.

Orders and medal bars are not put on under combat conditions; furthermore, all otherwise reflective or glossy uniform accessories or equipage are not worn in the field.

Special Distinctions

From the above universal descriptions, the following exceptions are applied to the waffenrock.

Generals.

On the field gray waffenrock of generals, on each sides of the chest, approximately 1cm above the height of the fourth buttonhole, are breast pockets with scalloped pocket flaps.

On the collar are 10cm long, poppy red patches, with gold old Prussian embroidery.

The cuffs on the field blouse of generals have a height from 16 to 19cm and have poppy red piping around the top only.

The coattail pocket flaps are provided with red piping, strait cut as they are on the frock coat and have a length of approximately 23cm.

General-Adjutants and Generals a` la suite of His Majesty the Kaiser.

The collar patches of the general's field blouse have a reduced size copy of the kolben embroidery as on the dark blue waffenrock, instead of old Prussian embroidery. The embroidery of adjutant generals is gold, for generals a` la suite it is silver. The buttons are correspondingly gold or silver.

Aiguillettes and fangschnüre[4] are worn as before.

Flügel Adjutants[5] of His Majesty the Kaiser

The field gray waffenrock is the same as it is for infantry officers, however cuffs and collar patches are as on the general's field blouse, the later has kolben embroidery. The buttons are silvered. Aiguillettes and fangschnure[4] are worn as before.

Personal Adjutants of Princes and Royal Houses

Field blouse as for infantry officers, however cuffs and collar patches are as on the general's field blouse, the collar patches however have triple silver embroidered braid edging. Silvered buttons. Aiguillettes and fangschnure[4] are worn as before.

Officers of the War Ministry and the General Staff

Field blouse as for infantry officers, however with Swedish cuffs and carmine red piping. The carmine red collar patches are without litzen. The buttons for officers of the war ministry are gilded, and those of the general staff silvered, respectively.

---

[4] Fangschnure: two round cords with metal tips or tassels hanging from helmet or shoulder.

[5] aide-de-camp

Cuirassiers and Jägers zu Pferde

The stand collar on the field blouse of the cuirassiers has field gray patterned braid, it consists of two narrow side stripes in the facing color of the regiment on the edges. For the jägers zu pferde there is light green patterned braid with two side stripes in the regimental color.

Collars and cuffs of the cuirassiers have no colored piping.

Dragoons

For dragoon regiments Nr. 11 and 12 the piping on the officer's field blouse is of carmine red velvet.

All Black Piping etc. on officer's field blouses is of velvet.

Fusilier Regiment Nr. 73, Infantry Regiment Nr. 79 and Jäger Battalion Nr. 10

The same "Gibraltar" cuff title worn on the dark blue (dark green) uniform is worn on the field blouse.

Officers of the Technical Institutes and the Cadets Corps

The field blouse is the same as the infantry officers, however with Swedish cuffs and gold-plated buttons. The officers of the cadet's corps have collar patches the color of the litewka collar patches. Single litzen, except the ones from the main cadet institute Gross Lichterfelde which have double litzen; furthermore, litzen is worn on the cuffs.

Mounted Field Jäger Corps

They have the same field uniform as officers of the 1st Guard Dragoon Regiment however, with dark green collar patches and no litzen on the cuffs.

Officers of the Clothing Office and the District Commands

Field blouse as it is for infantry officers with Brandenburg cuffs. Gilded buttons.

Medical Officers

Field blouse is as it is for infantry officers, however with Swedish cuffs. The patches on the collar are of dark blue cloth and have poppy red piping. Collar patches and cuffs are without litzen. The piping around the collar and cuffs is dark blue, however down the front and on the coattail pocket flaps the piping is poppy red.

Medical Officers with Generals Rank

Have the field blouse as described for generals, however in place of the poppy red collar patches with old Prussian embroidery, the patches are from dark blue cloth with poppy red piping and with a reduced copy of the gold litzen embroidery on the dark blue waffenrock.

Veterinary Officers

Field blouse as for medical officers, however the collar patches are black and have carmine red piping. Collar patches and cuffs are without litzen. On the cuffs and collar the piping is black, however down the front of the blouse and on the coattail pocket flaps it is carmine red.

Army Uniform

For the foot troops, the officers wear the same waffenrock as the line infantry with Swedish cuffs and gilded buttons.

For the mounted troops, the officers have the

waffenrock of the line dragoon regiments, the piping on the collar and the cuffs is poppy red. Down the front of the blouse and around the under border of the skirt as well as on the coattail pocket flaps, the piping is white.

Reserve Officers

Wear the same field uniform as the active officers in their units.

Landwehr (Militia) Officers

The officers of the Guard landwehr infantry and provisional landwehr infantry wear the uniform of officers of the previously mentioned Guard regiments and line infantry, respectively.

The officers of the Guard landwehr cavalry have, on the field gray waffenrock, gilded buttons and poppy red piping down the front and on the coattail pocket flaps. The waffenrock has a stand collar without piping (as it is on the cuirassiers). The poppy red collar patches on the waffenrock have one litze out of matte silver fabric (as it is for the Guard du Corps and Guard Cuirassiers). The cuffs are of the same form as the uhlans and have one silver litze, however on these cuffs there is no piping. Around the collar, as well as along the upper edge and down the back slit of the cuffs is a trim of field gray, patterned braid. On both sides are 2mm wide stripes of poppy red silk.

The officers of the provisional landwehr cavalry have the same waffenrock as the Guard landwehr cavalry, however without litzen on the collar and on the cuffs. In addition, along the lower edge of the flat braid trimming, there is a 2mm wide field gray soutache cord braid. In the upper corners and the rear of the collar as well as in the upper corners of the cuffs the cord braid makes loops, and under the cuff button makes a cord knot.

Armory and Ordinance Officers

As for officers of the line foot artillery, however Swedish cuffs are worn on the field blouse and the piping is black velvet throughout. Shoulder pieces as on the dark blue waffenrock. In the field, spurs are generally worn.

Fortress Construction Officers

As for engineer officers, however without collar patches on the field blouse and the piping throughout is of black velvet. Shoulder pieces are as on the dark blue waffenrock.

Military Officials

Waffenrock (field blouse) with stand and fall collar as it is for infantry officers. Swedish cuffs of the waffenrock fabric. Piping around the collar and the cuffs is in the fabric and color of the collar and cuff fabric on the dark blue waffenrock*)[6]. The shoulder pieces are attached as before but there are no epaulette holders.

Officers matt crown buttons gilded or silvered. All senior officials that have embroidery on the dark blue waffenrock, have on the field blouse, collar patches of dark blue cloth with piping as it is on the cap crown. The shoulder pieces are the same as those on the dark blue waffenrock.

Field cap. Of field gray cloth or doeskin as for

---

[6] The information given about the branch colors and about the other distinctive characteristics on the blue waffenrock for the military officials' category are in the publication: "The Uniform of the German Army Volume 1." Moritz Ruhl. Details contained therein.

officers. Cap band, piping, cockades and heraldic eagle or heraldic shield as on the dark blue cap.

Deviations

The military superintendents when privy war counselors with the rank of counselor of the 1st class, wear the field blouse under the provisions for generals. However, the piping around the collar and around the cuffs, down the front and back on the coattail pocket flaps are as above, they also have collar patches of dark blue cloth with carmine red piping and reduced, silver litzen embroidery. Buttons are white silvered.

Sergeant Lieutenants (Feldwebelleutnants) for Reserve etc. Formations

They have the dress and insignia of Vizefeldwebels (infantry) or Vizewachtmeisters (cavalry and artillery), however with the proscribed shoulder pieces for lieutenants on the blouse and the overcoat. Head cover, knapsack, and bayonet together with the belt are as regulation for officers in their respective units.

The Ulanka (Lancer's Waffenrock)

The ulanka for officers is of the same cut as that for other ranks; special attention should be made to make it comfortable for underclothes to be worn underneath.

The fabric of the ulanka is broadcloth or doeskin of field gray color, the lining of the blouse is also field gray. The sleeve lining is light.

The ulanka has a stand collar of the same fabric. For the Guard uhlans it has corners and for others the upper edges are rounded; the closure only has two hooks and eyes. The Guard uhlans have colored collar patches with litzen on them.

The cuffs are of the same fabric with colored piping,

they are as before, pointed at the top and have a button under the point; for the Guard uhlans a litze is worn on the cuff.

On the field uniform only shoulder pieces are worn, never epaulettes, so the epaulette holders are not included on the ulanka.

The curved breast flap and the piping on the chest, the back and the sleeve seams are as on the previous ulanka.

Draw string and side hooks as prescribed for the existing ulanka.

The Attila (Hussar's Waffenrock)

The field gray attila is of the same cut as the former interim attila for hussar officers. However, it has a fold over collar, it is closed with only two hooks and eyes, made soft and may not be higher than 5cm.

The fabric is broadcloth or doeskin of field gray color.

The trimming on the field Attila is of gray wool, flat cord with black threads, 7mm wide and 2 1/2mm thick.

The rosettes and toggles are of gray wool cord with black threads.

The shoulder buttons, the lining, the side hooks, and the pocket flaps on the Attila are the same as prescribed above for the waffenrock.

The only deviation that should be noted is that for the 3rd Hussar Regiment the attila pockets are without cord trim as the other regiments are.

The Field Trousers

The fabric used for the field trousers is trikot of field gray color, gray green trikot is prescribed for the field trousers of the jägers, schützen, machinegun units and jäger zu pferd (mounted jägers). Apart from the branches just mentioned the above instructions for the field trousers also apply to breeches. For field breeches whipcord fabric is also authorized.

The field trousers for generals and the adjutants, of His Majesty the Kaiser and the ruling princes, have stripes 4cm wide of poppy red cloth the same as before. Officers of the war ministry and the general staff have carmine red stripes.

An exception to this, and important to note, is that the trimming on the field trousers for hussar officers is not of gold or silver braid but of a 1.6cm wide patterned field gray braid made of silk.

Furthermore, mounted seat reinforcing on the field trousers must always be out of cloth or leather in the color of the basic fabric.

The Field Cap

The field cap for officers is of field gray (or gray green respectively) cloth or doeskin. The cap band as well as the piping are of the same color as on the dark blue uniform. All black cap bands and piping for officers are of velvet, also the carmine red cap bands and piping on the field caps of the officers of the 11th and 12th dragoon regiments are velvet.

The field cap is soft worked, without stiffening on the side pieces, in front in the middle a stiffener can be attached (if it is used) to hold up the crown. The diameter of the crown should be approximately 6cm larger than the diameter of the head. The field cap has a black lacquered brim of flexible leather, as well as a black lacquered chinstrap with buckle device. The chinstrap is attached with two black lacquered, flat metal buttons with rounded rims.

The Shoulder Pieces

The shoulder pieces for officers on the field gray uniform are the same as those worn on the dark blue waffenrock, attila or ulanka. Generally, on these as well as those for officers of the cuirassier and hussar regiments the regimental number or cypher is applied. An exception to this is found only for the Regiment Guard du Corps, the Guard Cuirassier Regiment, the Life Guard Hussar Regiment, and the Life Hussar Regiment Nr.1.

Description of the Field Clothing of Military Clergymen

Protestant and catholic military clergymen service dress in field conditions (unless donning of robes or liturgical vestments is necessary) consists of:

Frock coat (also called official dress or cassock) of field gray cloth with stand collar, cuffs and one row of buttons, the knee is covered, around the collar, the cuffs and down the front is violet piping. Field gray trousers or breeches, overcoat, and cape of gray cloth. Soft field gray felt hat (similar to the German protection troops in the colonies) violet hat band, violet brim border, Reich and state cockades, and in between those is a white enamel cross. The field gray cap with violet cap band has cockades and cross the same as on the hat. Ankle boots and gaiters out of brown leather. Gloves of red brown color. Around the left upper arm is provided an arm band made of white silk with one violet stripe around the band and on both sides (above and below) two white 2.5cm wide (horizontal) stripes, in the center it has a red cross. The total width of armband is 10cm.

As a chaplain's insignia a chain hangs around the neck with a cross on the breast. For the catholic field chaplains, also on the march and in combat, the violet stole.

The Gray Field Uniform of the German Army
General
Colonel of the General Staff
Infantry Officer 21st Army Corps
Guard Infantry Officer
Jäger Officer
Guard Pioneer Officer
Pioneer Officer
Guard Machinegun Unit Officer

The Officer's Gray Field Uniform
R. Guard du Corps Officer
Cuirassier Officer
7Cuir. R
Dragoon Officer
12 Drag. R.
Hussar Officer
16 Hus. R.
Field Artillery Officer
Horse Jäger
2nd Reg.
Guard Supply Officer
Guard Lancer Officer

The Officer's Gray Field Uniform
Bavaria Inf. Life Reg. Officer
Bavarian Jäger Officer
Bavarian Uhlan Officer
Chevalier Officer 4. Chev. Reg.
Saxon Life Gren. Reg 100 Officer
Saxon Schutzen Reg 108 Officer
Saxon Foot Artillery Officer
Saxon Guard Rider Reg. Officer.

The Gray Field Uniform of Medical Officers etc. and Military Officiers
Paymaster
Ordnance Officer
Medical Officer
Veteranary Officer
Fortress Construction Officer
Supply Officials
Judge Advocate
Intendant Official Works on the War Council

Description of the Enlisted Man's Uniform

In accordance with the official regulations on the distinctive insignias of the

Field Uniform for troops of the German Army

Not only are detailed descriptions of the individual pieces of clothing given here, but also lithographic color prints. On them are detailed panels and special pictures which show the characteristics and distinctions of the field uniform, so it will be presented in a clear and easily understandable way.

The insignia for the different army corps, branches etc., correspond in color almost always to the insignia on the previous peacetime uniform. The only exceptions are the two Saxon army corps (XII and XIX). The Saxon uhlans, artillery, pioneers, and supply train colors on the field uniform are the same as for the Prussian branches. The Saxon peacetime uniforms are entirely different.

The precise innovations are on the enclosed color prints for the Saxon units, and their emblems are shown to prevent any possible doubt about them.

The Waffenrock with Fold Collar

The waffenrock with fold collar, or stand and fall collar, is prescribed for foot troops, for the machinegun units, for the artillery and for the supply train formations; for infantry, pioneer, transportation troops, field and foot artillery and the supply train it is of field gray cloth, for jägers, schutzen and machinegun units it's of gray green cloth.

The coat must be large enough so that it can be worn over the drill jacket. On each coattail there is a curved pocket flap. At the front below the waist on each side of the coat are slanted pockets, and they are closed with round cornered pocket flaps. On the back of the coat, near the waist, to bear the garrison belt, are two waist hooks of tombac or nickel silver. Inside the coat at waist height is a draw string.

The buttons on the waffenrock are flat round edge matt metal buttons with a stamped crown from tombac or nickel.[7] Down the front, the waffenrock has one row of eight buttons. On each back-pocket flap are three buttons. To close the front pocket flaps on each side of the coat below the waist is a slightly smaller similar button. Furthermore, on each shoulder is a button with the company number etc. to button the shoulder straps.

The colored piping on the waffenrock, down the front, at the back and on the coattail pocket flaps is:

1. For infantry, machinegun units, pioneer, transportation troops, field and foot artillery and supply train, poppy red,
2. For jägers and schützen light green.

The fold collar of the waffenrock, on its outer and inner parts, is of the same field gray or gray green cloth as the waffenrock. It is closed with a black lacquered hook and eye. The collar on the front and around the bottom edge has piping of different colors.

The color of the piping on the fold collar is:

1. For the infantry, the 1st Guard and Nos. 1 to 8 Machinegun Units poppy red,
2. For jägers light green,
3. For schützen, 2nd Guard Machinegun Unit, artillery, pioneers, and transportation troops black,
4. For supply train light blue,

The cuffs are the same cloth as the waffenrock and have around the top and down the back slit colored piping. For the cuffs with litzen the piping is also around the lower edge of the cuff, (the lower edging was later abandoned, as early as 1912 in Bavaria).

The Swedish cuffs have on the outside of the sleeve two buttons set in a row along the upper edge.

The Brandenburg cuffs have patches also of the

---

[7] The matt uniform buttons as well as the side hooks, during the war, will be made galvanized from Siemens-Martin-Stahlblech.

same cloth. For the cuffs without litzen, as well as on the scalloped cuff patches of the Guard Schützen and the 2nd Guard Machinegun Unit the upper, back, and lower edges of the cuff patches have piping, the fourth side of the patch is always without piping.

The color of the piping on the cuffs is the same as the color of the piping on the collar.

The litzen are made of either white or yellow camel yarn and have in the middle a colored flash. In the accompanying table No. 20, for the field uniform, the prescribed litzen are visually depicted one third of the actual size.

On the fold collar the litzen are worn on both sides as prescribed. For those units that have double litzen, non-commissioned officers wear only one single litze, and with it the non-commissioned officer's braid is worn around the collar.

On the Swedish cuffs of both arms are two vertical litzen side by side on the outside of the sleeve.

On the patches of the Brandenburg cuffs, on each patch, are three horizontal litzen sewn on one above the other.

The shoulder straps have been up till now made removable and consist of an over strap and tongue, both of field grey cloth.[8] They are now sewn on to the shoulder seams.

The shoulder straps have colored piping and the emblems of the various branches or units. In the middle of the shoulder strap is the number or the monogram of the concerned unit and is universally red in color, while **the color of the piping on the shoulder straps** is for the army corps of the respective unit or the concerned branch hereafter specified.

---

[8] An exception has been made to this in the field uniform of medical workers (Krankentrager) and the military medical orderlies (Krankenwarter) etc. For the former they have shoulder straps of carmine red and the latter of dark blue cloth; therein the number of the respective Army Corps is in yellow.

White Shoulder Strap Piping for:

1st and 5th Guard Regiment of Foot,
1st and 5th Guard Grenadier Regiment,
The Infantry Regiments of the I, II, IX, X, XII Army Corps and the infantry regiments. 109,110,116.
The non-commissioned officer's schools at Potsdam and Ettlingen,
The Field Artillery Regiments of the I, II, IX, X, XII Army Corps,
All Foot Artillery Regiments,
The Instructor Battalion of the Foot Artillery Gunnery School,
The Test Battery of the Artillery Testing Commission.

Red Shoulder Strap Piping for:

2nd Guard Regiment of Foot
2nd Guard Grenadier Regiment
The Infantry Regiments of the III, IV, XI, XIII, XV, XIX, Army Corps and the Infantry Regiment 89 II Battalion, 111, 115, 168, 169, 171, 172
All Machinegun Units,
The non-commissioned officer's schools at Northeim and Marienwerder,
1st and 2nd Guard Foot Artillery Regiment,
(The Shoulder Straps of the 1st Guard Field Artillery Regiment have on the inner border red piping, with a narrow white stripe on the outer.)
The Field Artillery Regiments of the III, IV, XI, XIII, XIV, XV, XIX. Army Corps, the Field Artillery Regiments 25, 61,
The Instructor Regiment of the Field Artillery Gunnery School,
All Pioneer Battalions.

Yellow Shoulder Strap Piping for:

3rd Guard Regiment of Foot,
Guard Fusilier Regiment,
3rd Guard Grenadier Regiment,

The Infantry Regiments of the V, VI, XVI, XVII Army Corps and the Infantry Regiments 112, 118, 142
The non-commissioned officer schools at Wetzlar and Treptow,
The 3rd Guard Field Artillery Regiment
The Field Artillery Regiments of the V, VI, XVI, XVII Army Corps.

Light Blue Shoulder Strap Piping for:
4th Guard Regiment of Foot,
4th Guard Grenadier Regiment
The Infantry Regiments of the VII, VIII, XVIII, XX Army Corps
All Supply Units

Light Green Shoulder Strap Piping for:
The Infantry Regiments of the XXI Army Corps and the Infantry Regiment 114,
All Jäger Battalions
The Guard Schützen Battalion,
The Schützen (Fusilier) Regiment 108,
The Field Artillery Regiments of the XXI Army Corps.

Light Gray Shoulder Strap Piping for:
All Transportation Troops.

The **schwalbennesters**[9] are of field gray cloth. They use hooks to attach to the top of the sleeve at the shoulder seam. The braid is the same for musicians as for non-commissioned officers in the unit. For the foot troops it consists of eight vertical braids or trims, as opposed to those for trumpeters of the mounted troops which consists of eight braids sloping forward from the back top to bottom diagonally. Around the bottom of the schwalbennester is a strip of the same braid or trim. Under the latter and parallel with it is (with the exception of troop units referred to below) is a colored piping the

[9] "Swallows Nest", Shoulder Pieces for Musicians.

same color as the piping down the front and on the coattail pocket flaps of the field waffenrock.

On the schwalbennesters of the following units: The Guard Infantry Regiments, the Guard jäger and schützen, the pioneer, transportation troops and the foot artillery of the Guard, and Infantry Regiments Nos. 89, 109 and 115, there is no piping and instead there is a 3cm long thin fringe the color and material of the braid trimming. The braid for the musicians (hoboisten) and for the trumpeters is of the same pattern as the non-commissioned officer braid of the respective units. For the drummers (tambouren) and horn players (hornisten) of the foot troops the trimming on the schwalbennester is white half woolen braid. However, for those units who have yellow litzen on the waffenrock the braid is yellow.

The rank insignia for non-commissioned officers is braid 8mm in width. On the fall collar it runs along the lower edge and the front of the collar. On the Swedish cuffs the braid is along the upper edge and the rear side slit. For the Brandenburg cuffs along the upper edge of cuff and the lower front edge of the cuff flap. The double braid of the regimental sergeant-major[10] runs about 7mm above the cuff and around the sleeve but on the Brandenburg cuff, it is let in under the cuff flap, so the cuff flap is plain.

The Waffenrock with Stand Collar

The waffenrock with stand collar is for cuirassiers, dragoons and the jäger zu pferde. For cuirassiers and dragoons, it is of field gray cloth, for jäger zu pferde the cloth is gray green. Regarding the nature of the coat itself, it follows the previous description of the waffenrock with fold collar. The matt **metal buttons**, of which eight are down the front and three on each coattail pocket flap, are low domed of tombac or nickel with stamped crown.

However, the **piping on the waffenrock** is different

[10] Etatsmassigen Feldwebel or Etatsmassigen Wachtmeister.

because it is not only down the front and on the coattail pocket flaps but also runs around the bottom seam of the coat skirt.

The color of this piping is:

1. For cuirassiers and dragoons, generally it is the color of the regiment.
2. For the jägers zu pferde however, throughout it is light green.

The stand collar of the waffenrock has a height from 4 to 5 cm and is at the front upper corner either square or rounded and held closed with two or three black lacquered hooks and eyes. For the cuirassiers and dragoons the stand collar is of field gray cloth. For the jäger zu pferde and the stabsordennanzen it is of light green cloth.

The piping around the stand collar of the cuirassiers and dragoons, in general, is the color of the regiment. There is a deviation from this only for dragoon regiments 13, 14, 15 and 16, which have white collar piping, also Dragoon Regiment 22, has instead black collar piping. For the jägers zu pferde the piping on the stand collar of all regiments is light green.

The waffenrock with stand collar has **Swedish cuffs**. For the cuirassiers and dragoons, they are of field gray cloth and have, around the top as well as down the rear slit, piping in the regimental color. There is an exception for the above-mentioned dragoon regiments that have stand collars with white or black piping, the cuffs of those have the same color piping, respectively. For cuffs with litzen there is also piping around the lower edge of the cuff. For the jägers zu pferde and the stabordennanzen the cuffs are of light green cloth and have no colored piping.

Around the stand collar and the cuffs of the waffenrock of the cuirassiers, jägers zu pferde and stabsordennanzen is a **braid trimming**. For other ranks of camel yarn, for officers of silk.

For cuirassiers, the braid is field gray with white outer stripes and colored inner stripes, the later in the color of the regiment. For the 3rd Cuirassier Regiment, the braid has a special pattern.

The braid for the jägers zu pferde is light green and has in the middle a wide stripe in the color of the regiment, on both sides the stripe has a narrow yellow border.

The braid for stabsordennanzen is throughout lemon yellow with light green stripes.

Around the stand collar the braid trimming runs along the top and front edge and on the cuffs along the top edge and down the back slit. On the accompanying plates 18 and 19 are the colored braid trimmings at half the original size. Pictured in addition to those of the troops is the officer's braid since the latter diverges substantially from the former.

The **litzen** of the various units are displayed on plate 20. The non-commissioned officers have, if the unit has double litzen, only one single litze on the stand collar below the braid. For the cuirassiers however throughout only one single litze is worn below the braid trim.

On the Swedish cuffs, there are also two vertical litzen side by side on the outside of the cuffs.

The **shoulder straps** are of field gray cloth for the cuirassiers and dragoons and were removable as they were previously described for the fall collar waffenrock with an upper part and tongue. However, they are also sewn at the shoulder seam. For the jägers zu pferde and stabsordennanzen the shoulder straps are of light green cloth and, as previously, sewn at the shoulder seam.

The **piping on the shoulder straps** is always the distinguishing color of the regiment. For the cuirassiers, the shoulder straps have an inner narrow colored piping of white cloth as well. The insignia (regimental number or emblem) on the shoulder straps is red, an exception to this is only the Kings Regiment of Jäger zu Pferde Nr.1[11] which has a yellow shoulder strap badge. The stabsordennanzen light green shoulder straps (without piping) have the brass Roman numerals of their army corps.

On the accompanying plates the colors of the

[11] Regiment Königsjäger zu Pferd

shoulder straps are presented. Since several regiments have the same color shoulder straps, in those figures only the number or the emblem of one of the regiments listed under it will be shown. A precise compilation of all emblems is contained in eight wall plates of the German field uniform, drawn by Major Schmidt released by this same publisher (Moritz Ruhl). Format: 42x52 cm.

The **schwalbennesters** of the trumpeters are worn on the waffenrock of the cuirassiers and the dragoons. As already described, they are of field gray cloth and firmly hooked at the shoulder seem while those of the jäger zu pferde are of light green cloth and sewn to the shoulder seam. The braid trimming on the Schwalbennesters, as with all mounted troops, has seven strips of braid going diagonally from behind on top forward to the bottom. Along the lower edge is a strip of the same braid. For Cuirassiers and Dragoons there is a colored piping on the bottom edge and for jäger zu pferde the lower edge has no piping only a narrow strip of backing cloth.

The **rank insignia for non-commissioned officers** is the appropriate braid running around the upper edge and the front of the stand collar. For the cuirassiers, jäger zu pferde and stabsordennanzen the non-commissioned officer braid trimming is set in place of the regimental-colored braid, both on the stand collar as well as on the cuffs. In the case of the cuffs, the braid runs along the upper edge and the rear slit.

The Ulanka

The **ulanka**[12] is made of field gray cloth. It is double breasted, has curved breast flaps with piping and is buttoned from the left to the right side. There are seven buttons on the opening side, of which one is on the upper chest halfway between the neck and shoulder and six others that button to the side along the breast flap opening. There are seven (blind) buttons on the left chest

[12] Lancer tunic.

opposite the buttons on the right. On the ulanka they are matte **metal buttons** of tombac or nickel, low domed and crown stamped.

The **cuffs** on the ulanka are of field gray cloth. They rise on the outside sleeve side and form a point. Around the upper seam and the back slit is colored piping. For the ulan regiments that have litzen there is the same piping at the lower edge of the cuffs.

The rest of the ulanka is the same as the waffenrock of the Dragoons etc. also with stand collars including the pockets etc.

The **piping** on the ulanka runs around the stand collar and along the right breast flap edge as well as around the bottom of the skirt. On the left breast side is piping mirroring the right breast edge outline. Furthermore, the ulanka has piping on the back seams and along the back-sleeve seams that meets at the top. There is also piping on the curved sides of the coattail pocket flaps.

The **color of the piping** on the breast flap, on the left breast side, around the bottom skirt seam, on the sleeve and back seams as well as on the coattail pocket flaps is the regimental distinguishing color (also the color of the collar etc. on the peacetime uniform). Only for Ulan Regiments 12 and 16 is there an exception, for them the piping is white. However, the piping on the cuffs and around the stand collar for these two regiments is the regimental distinguishing color as it is for all other ulan regiments,

The ulanka has **shoulder straps** that are made to be removable and are thus made with an over piece and tongue. At the base of the strap the outer edge is rounded, and it tapers to the tip so that at the upper end is approximately only half as wide as the lower. The strap almost has a teardrop shape.[13] The cloth of the shoulder strap is field gray, in the middle is the regiment's number or distinguishing monogram in red color. Around the edge of the shoulder strap runs a piping of the same color

---

[13] For the Saxon Ulan Regiments 17,18 and 21 the shoulder straps are evenly 5.5 cm wide without rounded edges.

shown on the field of the epaulettes on the peacetime uniform (pictured on the accompanying plates).

For the **litzen** on the collar of the ulanka the same practice applies as what has already been said above for the waffenrock of the dragoons etc. On the cuffs of the ulanka there is but only one vertical litze on the outside of the sleeve.

Also, with respect to the **schwalbennesters** and affixing the **non-commissioned officer's braid** on the stand collar the same provisions apply as mentioned for the waffenrock of the dragoons etc. The non-commissioned officer's braid on the tip of the cuff of the ulanka runs along the upper edge and the back slit.

The Attila

The cloth of the **Attila,** including the round cornered **stand collar** is field gray. The **cord braid** on the Attila is of field gray flat cord. There are five horizontal rows of double flat cord across each side of the chest. From top to bottom the rows go from longer to shorter. At the outside ends they make a rosette. There is also a braid of gray flat cord on the upper and lower edges of the collar, as well as the front edge of the coat and around the skirt seam. On the sleeve there is cord braid marking the cuffs. The braid goes from the cuff slit upwards and makes a point on the outside of the sleeve. Then it makes a cord knot above the point. On the backside of the Attila on both back seams as well as around the skirt is cord braid.

For the Life Guard Hussar Regiment, on the collar and on the outside cuff below the cord, is a trim of **field gray braid**, showing a narrow yellow inner and red outer stripe.

In the front at hip height, located on each side of the Attila, are two scalloped pocket flaps closed with black horn buttons.

On the rear of the Attila, sewn on the waist as on the waffenrock, are two black oxidized side hooks of tombac.

The closures of the Attila, located on the right front breast, are a row of five vertically placed **toggle buttons** of black horn. To improve the fastening of the Attila there are also three hooks and eyes available.

The colors of the peacetime hussar regiments' uniforms are also represented on the field attila. There are no shoulder straps, but distinctive colored shoulder cords, showing the peacetime (attila and cord trim) colors of the concerned regiments. These shoulder cords consist of two pieces of colored double flat cord connected, so that the shoulder piece is four cords creating a single flat cord. They form a top loop to button on to the shoulder button, and at the bottom they are sewn into the shoulder seam. For the Life Guard Hussar Regiment however the shoulder cord is different, here there are only two broad red and yellow distinctive flat cords.

On the shoulder cord is found the metal number or distinctive monograms of the concerned regiments, sliver plated or brass. An exception is made only for the Life Guard Hussar Regiment and for the 1st Life Hussar Regiment, they have neither a number nor distinctive monogram.

The Trumpeters of the Hussar regiments have schwalbennesters of field gray cloth, however without colored piping around the bottom edge. The braid trimming is the same as it is for other mounted units.

The non-commissioned officer's braid on the collar of the Attila is around the upper and front edge of the collar along the cord trimming and, on the sleeve, it runs under the cord trimming of the cuff.

The Field Cap

The **field cap (feldmütz)** is of the same field gray or gray green cloth resp.as the waffenrock, the ulanka or the attila. It has a colored cap band of facing cloth and colored piping around the crown. In addition, there are individual units with one or two, and some Hussar regiments with even three-color piping on the band of the cap.

The color of this piping as well as the cap band are

on the accompanying plates containing pictures of the field caps.

On the front in the middle of the cap band is the appropriate state cockade and above it on the gray cloth the black-white-red German cockade.

To conceal the cap band on the field cap of the infantry, machinegun units, pioneers, telegraph troops and the cavalry (except jäger zu pferde) for combat in the trenches a **gray band** about 3.2cm wide of adjustable length has been introduced. There is a slit in the gray band to stick through the cockade.

The field uniform also has the **service cap (dienstmutze)** which differs from the field cap in that it has a brim of flexible black leather and has a chin strap of blackened leather.

The Field Trousers

1. The **trousers (tuchhose)** for jäger, schützen, machinegun units as well as the jäger zu pferde and stabsordennanzen are gray green, for all other units they are of field gray cloth. They have colored piping. Except for the jäger, schützen the jäger zu pferde and stabsordennanzen who have light green piping, the piping on the trousers is of red badge color. However, the variations of colored piping on the long trousers to be observed are as follows: the 2nd Cuirassier Regiment, the 11th and 12th Dragoon Regiments, as well as the 10th and 14th Ulan Regiments have carmine red; the 5th Cuirassier Regiment, as well as the 3rd, 7th and 15th Dragoon Regiments have pink; the 5th Hussar Regiment deep red; and the 10th Hussar Regiment pompadour red; otherwise, all piping on the trousers is poppy red.

2. The **reithose**[14] for the jäger zu pferde and stabsordennanzen is of gray green and for all other units of field gray cloth. The riding trousers only have piping for the jäger zu pferde, stabsordennanzen and the machinegun units. For the first two, they are light green

---

[14] Riding Breeches

and for the latter, poppy red cloth.

The **reithose of the hussars** have braid trimming covering both outside side seams. A field gray woolen braid is prescribed but for the Life Guard Hussar Regiment there is a yellow outer and an inner red stripe.

3. **Stiefelhose**[15] of the machinegun units is of gray green cloth and has piping of poppy red badge cloth.

General Remarks

For the field uniform with the fall collar waffenrock a **neck cloth (halstuch)** of gray cotton is worn and for the stand collar waffenrock etc. a **neck stock (halsbinde)** of field gray cotton is worn.

The **leather gear**, including the footwear, are worn in an unblackened tanned condition during mobilization. The white leather gear must be blackened. The sabretache[16] of the Hussars, the chin strap on the service cap and the helmet, the leather pieces on the faustreimen[17], portepees etc. are blackened.

The **metal pieces** on clothing and equipment in the field should not be brightly polished.

On the **field gray helmet etc. cover** all units shall bear a green number (except for the Guard Corps).

For the **drums**, the drum strips must also be painted gray for use in the field and the drum shell must be covered with helmet-cover fabric.

The **overcoat** in the field will have the same shoulder straps as on the waffenrock etc. An exception to this is made only for the jäger zu pferde and for the stabsordennanzen.

The **insignia of the medical troops,** worn with the same field uniform of their units, consists of a staff of Aesculapius in lemon yellow cloth on an oval field of gray or gray green cloth, respectively. It will only be sewn on during mobilization, in the middle of the right upper arm.

---

15 Boot Breeches

16 Säbeltasche

17 See German glossary of terms.

When commencing or during mobilization an offizierstellvertreter[18] wears the uniform and equipment of a vizefeldwebel or vizewachtmeister with the following deviations:

1. The officer's bayonet[19] will be carried on the prescribed officers under belt. Those in the mobilized infantry will carry the officer's knapsack (offiziertornister).

2. The insignia of the offizierstellvertreter is as stated:

   a. On the waffenrock and overcoat (except ulans) the shoulder straps have the non-commissioned officer's braid of the unit around the outer edge, on the point and the long sides. The number or distinguishing monogram etc. are how they are on the shoulder pieces of Lieutenants.

   b. On the Attila around the edge of the shoulder cord is an 8.5 mm wide double braid in the color and pattern of the non-commissioned officer's braid of the unit. The number etc. is how it is described under a.

   c. On the ulanka and on the overcoat of the uhlans around the edge of the shoulder straps is the non-commissioned officer's braid. The number etc. as under a.

   d. For the Landsturm Formations if they have litewken instead the waffenrock, the shoulder straps are edged as stated in a and have the number of the collar patches.
   Furthermore, on the overcoat are two stripes of non-commissioned officer's braid on the collar patches.

The braid edging under a, b, c and d is a visible 1mm wide border on the shoulder strap or shoulder cord, one exception is the Life Guard Hussar Regiment, for them the braid edging on the border of the shoulder cord runs completely around the cord.

---

[18] A non-commissioned officer in an officer's position.
[19] Offizierseitengewehr

Overview of the Gray Field Uniform in the Bavarian Army

The description of the field uniform of enlisted men of the Army of the Kingdom of Bavaria is essentially in accordance with the previous descriptions and the provisions contained on previous pages, while the remaining major deviations are indicated below:

The waffenrock with fall collar for the infantry, jäger, machinegun units, pioneers, transportation troops, artillery, and supply train, is throughout of field gray cloth. The exception is the field uniform of jäger and machinegun units, which for other contingents is of gray green cloth, in Bavaria it is not.

The piping for the Bavarian jäger battalions and for the Bavarian machinegun units is light green.

All buttons on the field uniform of Bavarian units are stamped with the heraldic Bavarian lion.

The waffenrock with fall collar and neck stock for the schwere reiters only has red piping down the front and on the back-pocket flaps (not around the lower skirt edge). The cuffs on these waffenrocks are of the Swedish form with two standing lion buttons placed horizontally side by side. The field gray shoulder straps have piping of the (regimental) facing cloth and a red regimental number. On the inner side of the fold collar a cord is sewn on. The neck stock is fastened by means of loops.

The double breasted waffenrock with fall collar and neck stock for chevaulegers and uhlans has two breast flaps with colored piping on the front edge. Both the right and the left breast flap can be worn outside, the latter case is the usual. On these waffenrocks, in addition to the back-skirt pocket flaps, all coat edges have piping in the regimental facing color. For the uhlans, such piping also runs along both back and sleeve seams.

The cuffs and shoulder straps for the chevaulegers are the same as for the schwere reiters. The Bavarian uhlans have the same pointed cuffs and the same rounded shoulder straps as the Prussian uhlan regiments. On the inner side of the fall collar is a neck stock attached in the same way as the schwere reiters.

The field gray cloth trousers have red piping. Only the jäger battalions and the machinegun units have light green piping. On the cloth trousers of the cavalry regiments, which are only worn by the tradesmen of the reserve squadrons, there is piping in the facing color of the concerned regiments.

The riding breeches[20]of mounted troops are without piping. Only machinegun units have light green piping on both the riding breeches and the boot breeches.[21]

The Uniform of the Landsturm Formations

The Landsturm Formations have a waxcloth cap with the landwehr cross of yellow metal. These caps are now made of dull field gray wax cloth.

Instead of the waffenrock, the litewka[22] is usually worn. The collar patches, (for the infantry poppy red, for the pioneers black with poppy red piping, for the field artillery black and for the foot artillery black with lemon yellow piping), will in future be discontinued. Also, the shoulder straps are of colored webbing and contain no number. All the concerned branches are identified by their different colors.

The Colors of the Shoulder Straps are for:

The Infantry = blue,
The Pioneers = black,
The Field Artillery = poppy red,
The Foot Artillery = yellow.

The same shoulder straps are also sewn on the overcoat. On the latter, as on the litewka, the collar

[20] Reithosen
[21] Stiefelhosen
[22] Undress jacket

patches - including the litzen - are discontinued.

The **Landsturm - Cavalry** wear the uniform of the established cavalry regiments with the landwehr cross.

The insignia for the identification of the individual Landsturm formations in the above-mentioned branches is worn on the front of the collar (both sides) of the waffenrock or the litewka and the overcoat as appropriate. It is the number of the Army Corps in roman numerals (for the Guard Corps a "G"), underneath that is the number of the battalion etc. in Arabic numerals within that Army Corps from number 1 (including Landsturm-Ersatz-Formations) marked in sequence. The numerals are of matt brass. However, it is permitted to use the existing bright Arabic numerals until stocks are used up.

For rapid mobilization, future Landsturm- Infantry-Battalions will have on the cover of the cap under the Landwehr cross only the battalion number.[23][24]

---

[23] The previous designation for all formations of the Landsturm will be retained during the present war. They are the same however, the new designations will be added in brackets e.g. Landsturm Infantry Battalion Recklinghausen (VII. 15), or 10th Landsturm Infantry Ersatz Battalion VII. Army Corps (VII. 42).

[24] During 1914 the Landsturm was reorganized, units formerly known only by area were now numbered by corps and battalion.

About the uniform adopted during mobilized conditions for contractors, that are not military service personnel.

The uniform for non-military civil doctors, contractually committed for service in mobile and non-mobile formations.

For these doctors, the following uniform is prescribed:

Gray jacket in the style of the litewka[25]. In place of the (medical officers) blue patches, on both sides of the collar is a staff of Aesculapius, no shoulder pieces or grade insignia. Around the arm is a white armband with a red cross. As headgear medical officer's wear, the cap. Long trousers or boot breeches are optional.

As weapons the officer's seitengewehr with portepee will be carried. Furthermore, the Mauser-self-loading pistol (7.63mm) is authorized.

The uniform for the following civilian contractors: pilots, observers, engineers as well as master craftsmen for the flying formations.

I. Nonmilitary Pilots and Observers.

Field gray service cap with visor and chinstrap as for the transportation troops.

The field blouse, trousers and overcoat are the same as for other ranks of the transportation troops, however without shoulder straps and litzen, and on the front edges of the coat collar is a winged propeller of gold metal, as it

---

[25] Undress fly-front jacket.

is on the shoulder pieces of officers of the flying troops. The short-edged seitengewehr M 98 with saber tassel and garrison belt will be carried as for the other ranks of the flying formations.

II. Those serving the flight troops engaged in engineering.

The uniform is the same as previously described under I above, however instead of the winged propeller are two crossed propellers.

III. The werkmeisters[26] of the flying formations.

A) Workshop personnel with actual responsibility as foremen wear the uniform of supervisors etc. in the transportation troops.

B) Workshop personnel appointed to management (deputy officials) wear the uniform of officials in the position they perform with the following exceptions:

1. In place of officer's shoulder pieces: they are replaced with light gray shoulder straps with no monograms except braid on both sides and along the outer edge following the pattern of the epaulette holder braid. The underlay of the coat shoulder strap is dark blue, the overcoat shoulder straps are of gray cloth.
2. In place of the gold portepee is a portepee of silver and black satin.
3. In place of the silver braid belt is an under belt of black patent (lacquered) leather.
4. Cap without the small silver eagle.
5. Helmet with split pin side buttons, helmet plate without the small official's eagle.

C) For non-military service technician contractors, the uniform is as described above under I,

---

[26] Equivalent to civilian foreman or deputy officials ranking as non-commissioned officers.

however without collar badges.

Uniform and Insignia of Airship Crew Personnel.

While in a mobilized status the following will be worn:

1. For contractual airship commanders and chief engineers in non-compulsory military service or in compulsory service without military training:

The field gray field cap has a black cap band with red piping as worn by officers of the airship troops.

A white armband with non-commissioned officers braid and the stamp of the service position and the imprinted insignia of the formation is worn.

2. For airship commanders and chief engineers in compulsory service with military training (if they are not previous officers entitled to officer's uniform) the uniform of a Feldwebelleutnant[27] of the airship troops is worn.

3. For contractual personnel without military training as well as compulsory airship helmsmen, airship radio operators and airship machinists without military training.

The field gray service cap with black, red piped cap band, provided with peak and chinstrap as for the airship troops.

A white armband as described under 1, however without non-commissioned officers braid.

4. For airship helmsmen, airship radio operators and airship machinists in compulsory service with military training, the enlisted uniform and rank corresponding to their position.

---

[27] Sergeant Lieutenant

Military Rank of Non-Military Compulsory Service Airship Personnel

Those not in military service positions but contractually obligated to the Army in the event of war as crew personnel of Reich's airships, as well as those of army commandeered private airships, are given military rank as follows for the wartime contracted period:

A) airship commanders and chief engineers the rank of officer and

B) airship helmsmen, airship radio operators and airship machinists the rank of feldwebel.

Furthermore:

C) airship commanders and chief engineers in compulsory service without officer rank, while in a mobilized status, are given the rank of officer.

D) airship helmsmen, airship radio operators and airship machinists in compulsory service without unteroffizer (non-commissioned officer's) rank, while in a mobilized status, are given the rank of unteroffizer.

Imperial Volunteer Automobile Corps during Mobilization

The clothing and equipping of the Imperial Volunteer Automobile Corps directed for all members and mechanics, in the field or in the occupation army while mobilized has been approved according to the Corps' existing clothing regulations.

Currently, the members of the Corps wear a waffenrock with a stand and fall collar of the same color cloth. Deep madder red collar patches without flat cord are authorized. Furthermore, they may wear in place of plain buttons matt silver buttons with raised stamped crown. The fangschnur need not be utilized.

In addition, for the mobilized force, the insignia from the clothing regulations of the Corps are as follows:

1) The chief of staff of the Corps wears field grade officer's shoulder pieces with piping according to the color on the collar without any insignia.

2) Members, that are not officers on leave of absence or retired officers, wear lieutenant's shoulder pieces with underlay in the color on the collar without any insignia.

3) The mechanics wear smooth silver braid around the collar.

The mechanics that are members of the Imperial Volunteer Automobile Corps used by the field and occupation army wear a white armband with an eagle on their upper left arm.

Military Rank of Members etc. of the Imperial Volunteer Automobile Corps

1) Members of the corps who are not on military service and who are contractually employed by the wartime field or occupation army as drivers with their passenger cars, are generally accorded military rank as officers for the duration of their mobile service.

2) Mechanics who are members of the corps and not in military service who are employed as mechanics for passenger cars in the event of war are generally accorded the military rank of unteroffiziere during mobilized service.

3) Furthermore, members of the corps in compulsory service who do not have the rank of an officer, during mobilization shall also receive the military rank of officer and

4) Members of the corps who are mechanics and who are on compulsory service similarly receive the military rank of unteroffiziere.

Volunteer Motorboat Corps during Mobilization

The clothing of members of Volunteer Motorboat Corps during wartime is also the general club dress.

Insignia is the German cockade on the cap, and worn around the arm, is a black, white, and red armband with a heraldic eagle.

The vehicle driver wears the shoulder pieces of lieutenants if they are not entitled to higher rank based on previous service.

As weapons members of Volunteer Motorboat Corps carry a blackened seitengewehr, over the jacket a belt and pistol are worn.

The vehicles of the Volunteer Motorboat Corps fly, at the stern, the national flag with heraldic eagle and on the front or on the mast a pennant similar to the Navy Jack[28], (black-white-red with heraldic eagle). The club pennant is not flown.

Military Rank of Members etc. of the Volunteer Motorboat Corps

1) The members of the Volunteer Motorboat Corps, in non-military service positions as contractors during wartime, who are skippers of their boats for the Army, receive for the duration of general mobilization the military rank of officers.

2) The boat helmsmen and boat machinists of the Volunteer Motorboat Corps in non-military service positions, receive for the duration of general mobilization "unteroffiziere" rank.

3) The compulsory service members of the Corps that have no officer rank, receive while mobilized, the rank of officer and

4) The compulsory service boat helmsmen and boat machinists in the same way receive the military rank of unteroffiziere.

5) The other crewmen of the boat are equivalent to privates in the land army.

---

[28] Gosch der Marine

II.

The Gray Field Uniform for Enlisted Men

1.

4. Guard Reg. of Foot Sergeant

5. Guard Gren. Reg.

Guard Supply Reg.

3. Guard Field Arty R.

Guard Jäger Batt.

Guard Pioneer Batt.

2. Guard Machinegun Unit

Guard Schutzen Batt.

The Gray Field Uniform for Enlisted Men

The Field Gray Uniform for Enlisted Men

Guard du Corps — Field Artillery I Army Corps — Life Guard Hussars R. — 5 Cuirassier R. — 6 Hussar R. — 13 Hussar R. — 7 Foot Artillery R. — 13 Pioneer Batt.

II. 4.

The Field Gray Uniform for Enlisted Men

3 Guard Ulan R. | 5 Ulan R. | 1 G. Drag. R. | 15 Drag. R | 16 Ulan R. | 6 Drag. R. | 9 Supply Train Batt. | 4 R. Horse Jägers

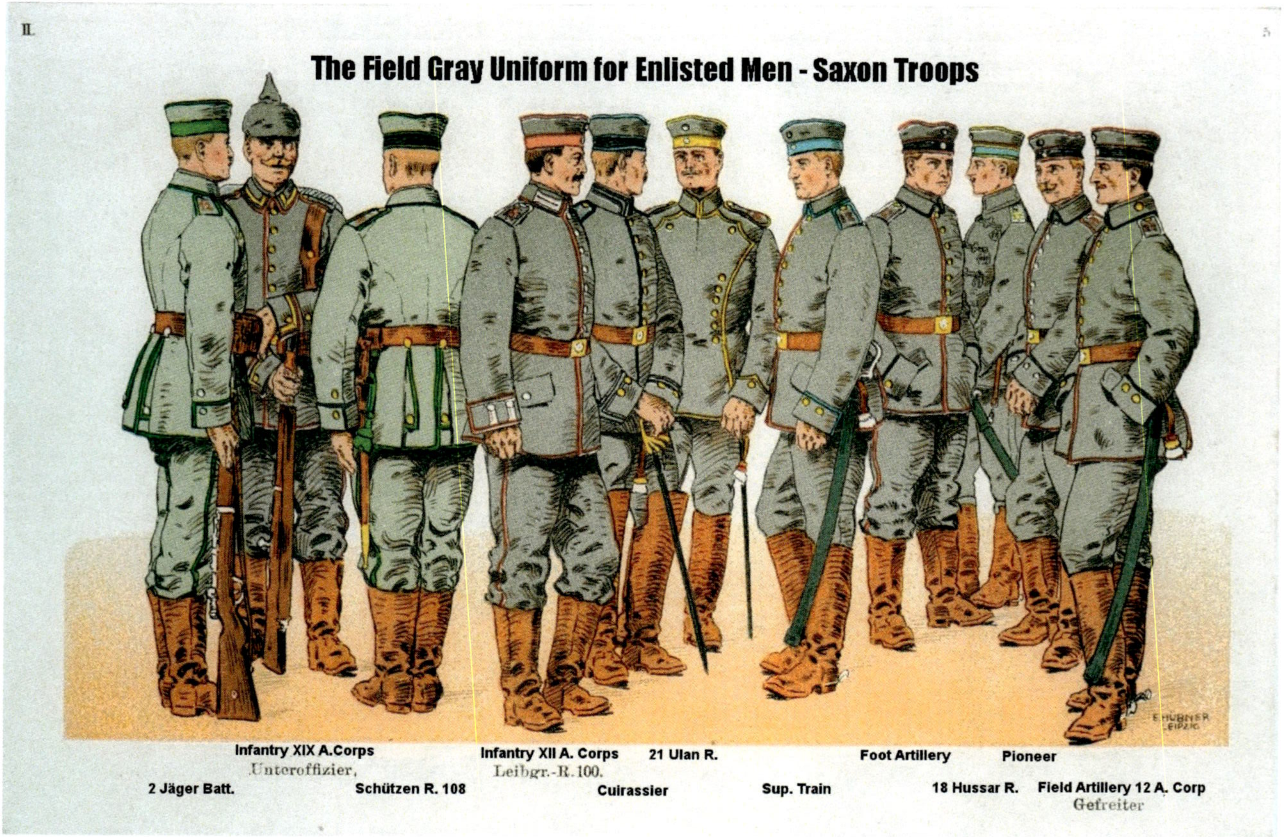
II.
The Field Gray Uniform for Enlisted Men - Saxon Troops
Infantry XIX A.Corps
Unteroffizier,
Infantry XII A. Corps
Leibgr.-R. 100.
21 Ulan R.
Foot Artillery
Pioneer
2 Jäger Batt.
Schützen R. 108
Cuirassier
Sup. Train
18 Hussar R.
Field Artillery 12 A. Corp
Gefreiter

II.
The Field Gray Uniform for Enlisted Men - Bavarian Troops
E. HÜBNER LEIPZIG
Jäger
Infantry
II A.C.
III A.C.
1 Ulan R.
Foot Artillery
Pioneer
1 Heavy Rider R.
Infantry Life R.
3 Chevauleger R.
Supply Train
Field Artillery

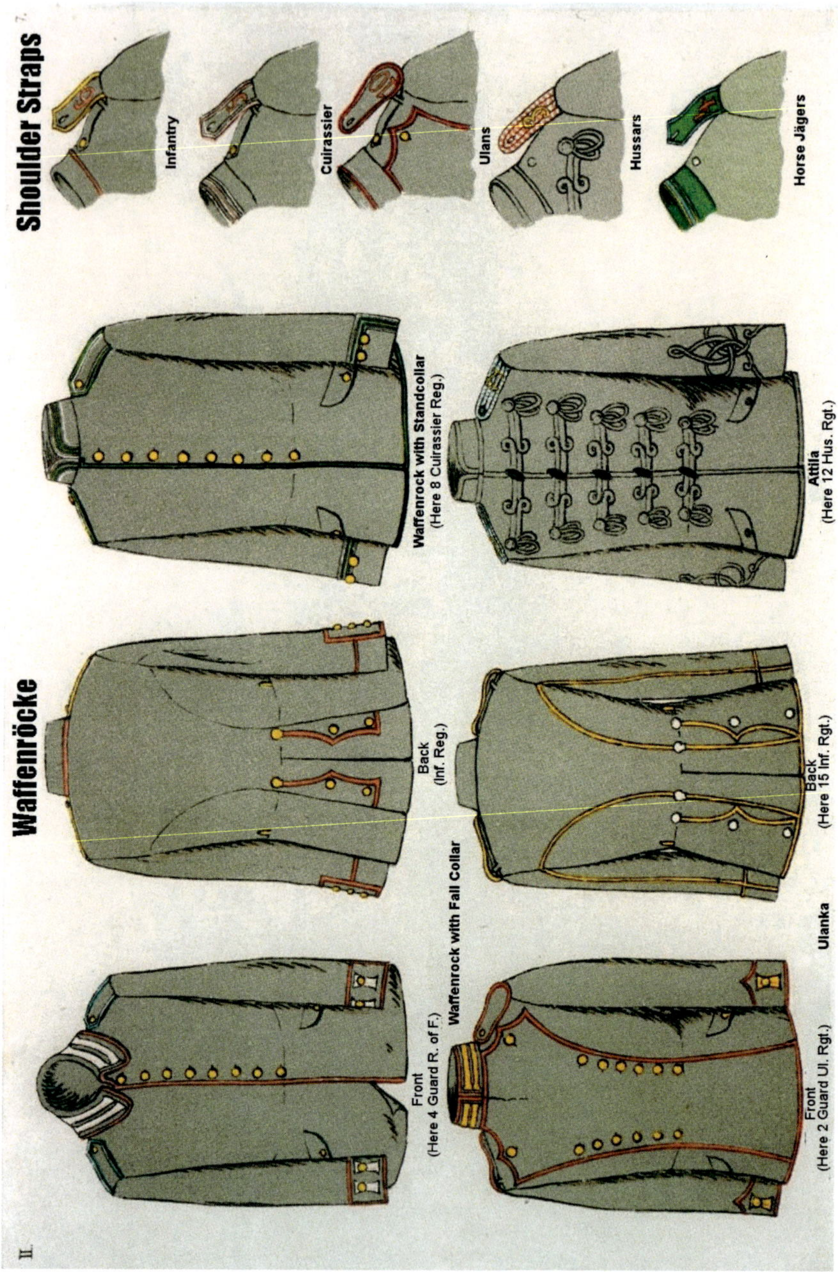
II.
Waffenröcke
Front
(Here 4 Guard R. of F.)
Waffenrock with Fall Collar
Back
(Inf. Reg.)
Waffenrock with Standcollar
(Here 8 Cuirassier Reg.)
Ulanka
Front
(Here 2 Guard Ul. Rgt.)
Back
(Here 15 Inf. Rgt.)
Attila
(Here 12 Hus. Rgt.)
Shoulder Straps
7.
Infantry
Cuirassier
Ulans
Hussars
Horse Jägers

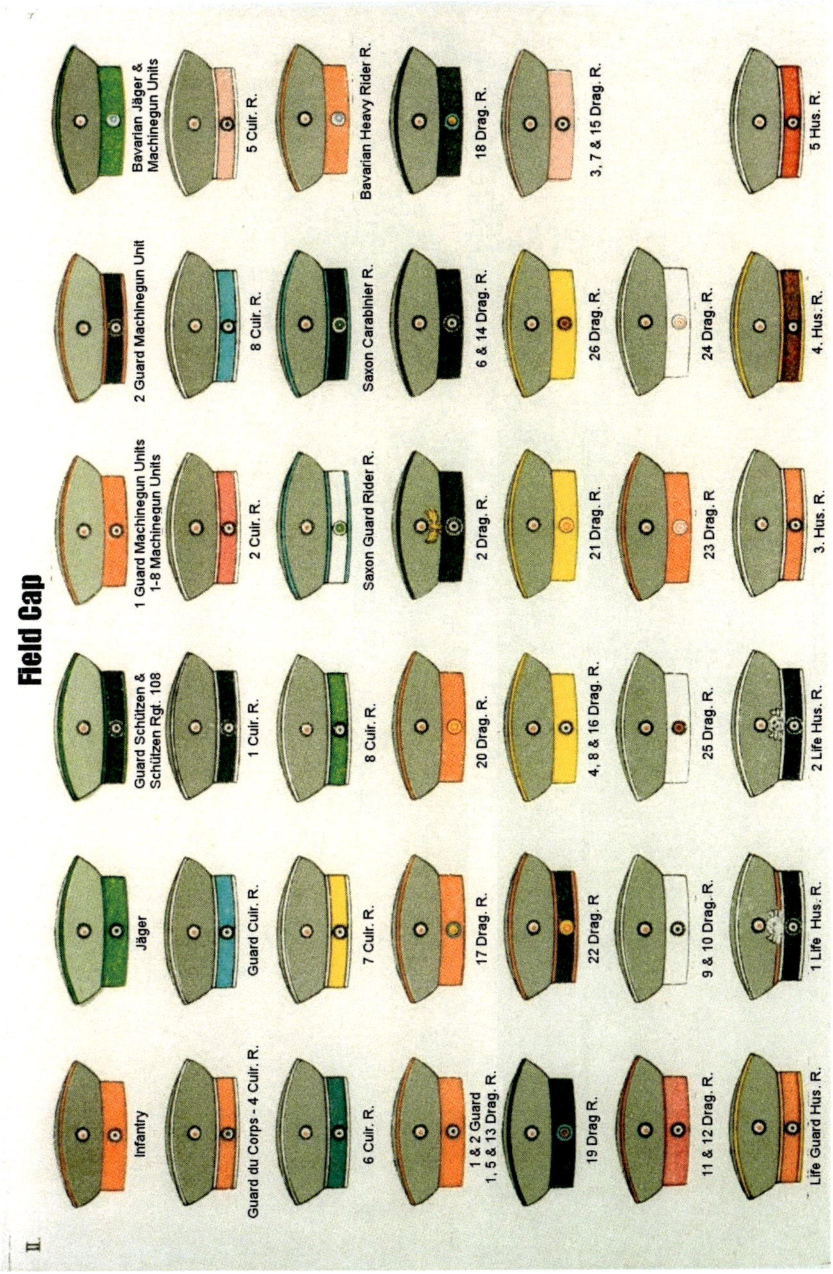
II
Field Cap
Infantry
Jäger
Guard Schützen & Schützen Rgt. 108
1 Guard Machinegun Units 1-8 Machinegun Units
2 Guard Machinegun Unit
Bavarian Jäger & Machinegun Units
Guard du Corps - 4 Cuir. R.
Guard Cuir. R.
1 Cuir. R.
2 Cuir. R.
8 Cuir. R.
5 Cuir. R.
6 Cuir. R.
7 Cuir. R.
8 Cuir. R.
Saxon Guard Rider R.
Saxon Carabinier R.
Bavarian Heavy Rider R.
1 & 2 Guard 1, 5 & 13 Drag. R.
17 Drag. R.
20 Drag. R.
2 Drag. R.
6 & 14 Drag. R.
18 Drag. R.
19 Drag R.
22 Drag. R
4, 8 & 16 Drag. R.
21 Drag. R.
26 Drag. R.
3, 7 & 15 Drag. R.
11 & 12 Drag. R.
9 & 10 Drag. R.
25 Drag. R.
23 Drag. R
24 Drag. R.
Life Guard Hus. R.
1 Life Hus. R.
2 Life Hus. R.
3. Hus. R.
4. Hus. R.
5 Hus. R.

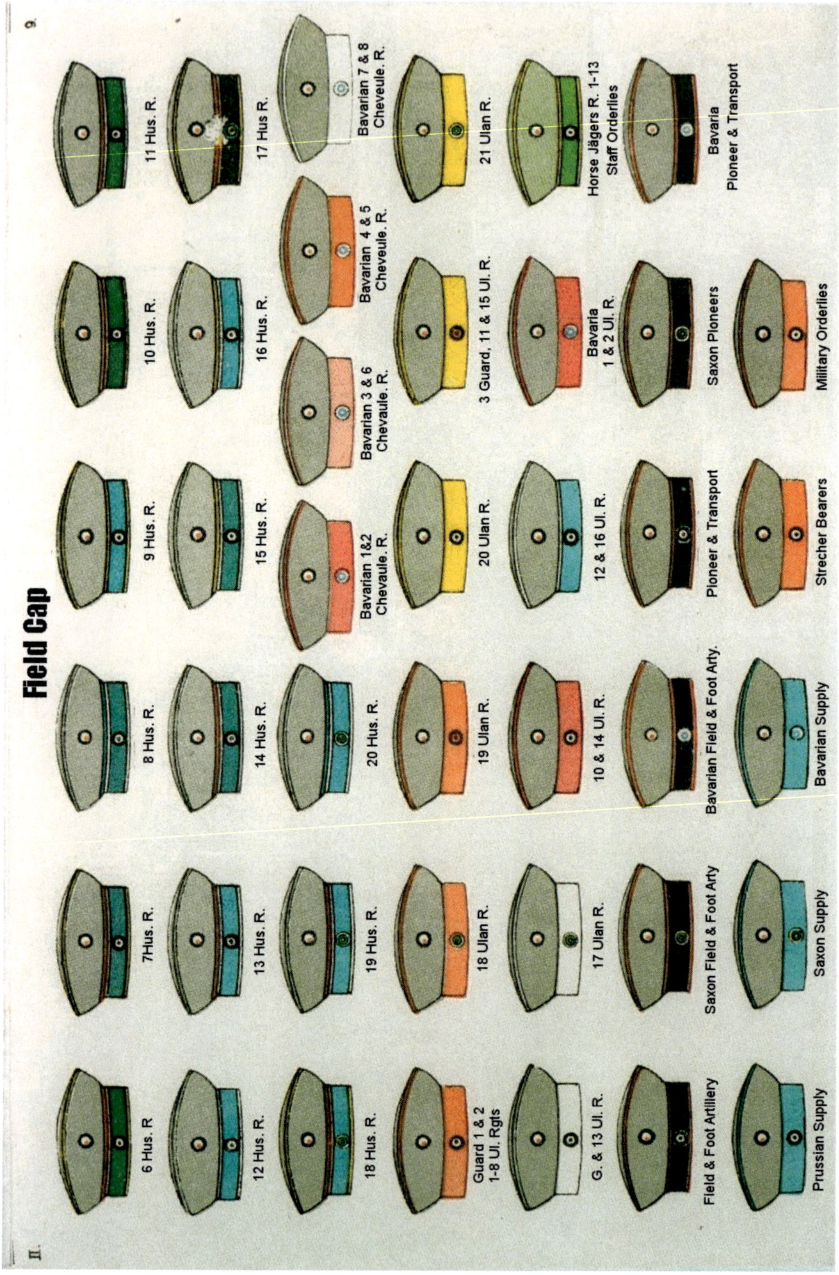
II.
9.
Field Cap
6 Hus. R
7Hus. R.
8 Hus. R.
9 Hus. R.
10 Hus. R.
11 Hus. R.
12 Hus. R.
13 Hus. R.
14 Hus. R.
15 Hus. R.
16 Hus. R.
17 Hus R.
18 Hus. R.
19 Hus. R.
20 Hus. R.
Bavarian 1&2 Chevaule. R.
Bavarian 3 & 6 Chevaule. R.
Bavarian 4 & 5 Cheveule. R.
Bavarian 7 & 8 Cheveule. R.
Guard 1 & 2 1-8 Ul. Rgts
18 Ulan R.
19 Ulan R.
20 Ulan R.
3 Guard, 11 & 15 Ul. R.
21 Ulan R.
G. & 13 Ul. R.
17 Ulan R.
10 & 14 Ul. R.
12 & 16 Ul. R.
Bavaria 1 & 2 Ul. R.
Horse Jägers R. 1-13 Staff Orderlies
Field & Foot Artillery
Saxon Field & Foot Arty
Bavarian Field & Foot Arty.
Pioneer & Transport
Saxon Pioneers
Bavaria Pioneer & Transport
Prussian Supply
Saxon Supply
Bavarian Supply
Strecher Bearers
Military Orderlies

II. 10.

Shoulder Straps

Of the shoulder straps shown, of course, only one number or a badge from the units listed in the subsections could be given.
The monograms shown are the same as those in Part I of the work: "The Uniforms of the German Army", but here they are red.

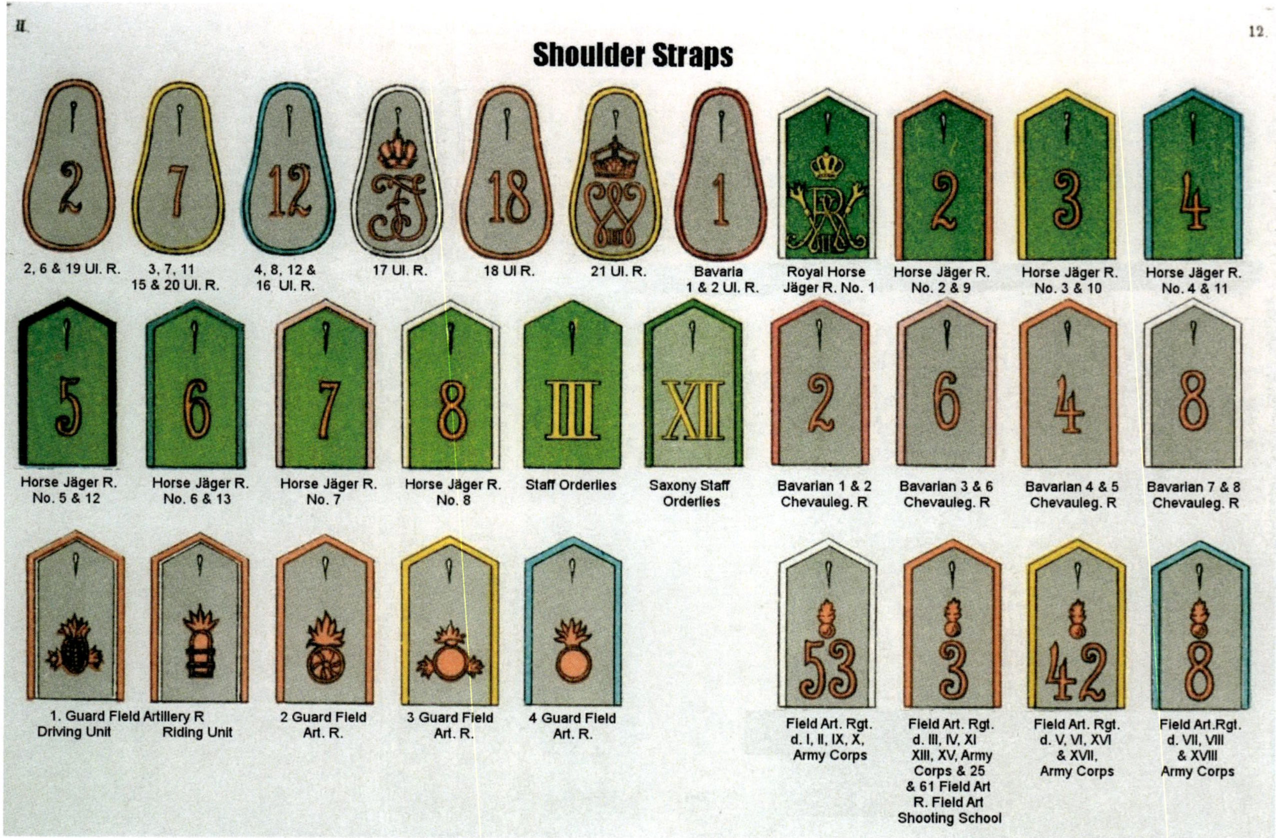
II.
12.
Shoulder Straps
2, 6 & 19 Ul. R.
3, 7, 11 15 & 20 Ul. R.
4, 8, 12 & 16 Ul. R.
17 Ul. R.
18 Ul R.
21 Ul. R.
Bavaria 1 & 2 Ul. R.
Royal Horse Jäger R. No. 1
Horse Jäger R. No. 2 & 9
Horse Jäger R. No. 3 & 10
Horse Jäger R. No. 4 & 11
Horse Jäger R. No. 5 & 12
Horse Jäger R. No. 6 & 13
Horse Jäger R. No. 7
Horse Jäger R. No. 8
Staff Orderlies
Saxony Staff Orderlies
Bavarian 1 & 2 Chevauleg. R
Bavarian 3 & 6 Chevauleg. R
Bavarian 4 & 5 Chevauleg. R
Bavarian 7 & 8 Chevauleg. R
1. Guard Field Artillery R
Driving Unit
Riding Unit
2 Guard Field Art. R.
3 Guard Field Art. R.
4 Guard Field Art. R.
Field Art. Rgt. d. I, II, IX, X, Army Corps
Field Art. Rgt. d. III, IV, XI XIII, XV, Army Corps & 25 & 61 Field Art R. Field Art Shooting School
Field Art. Rgt. d. V, VI, XVI & XVII, Army Corps
Field Art.Rgt. d. VII, VIII & XVIII Army Corps

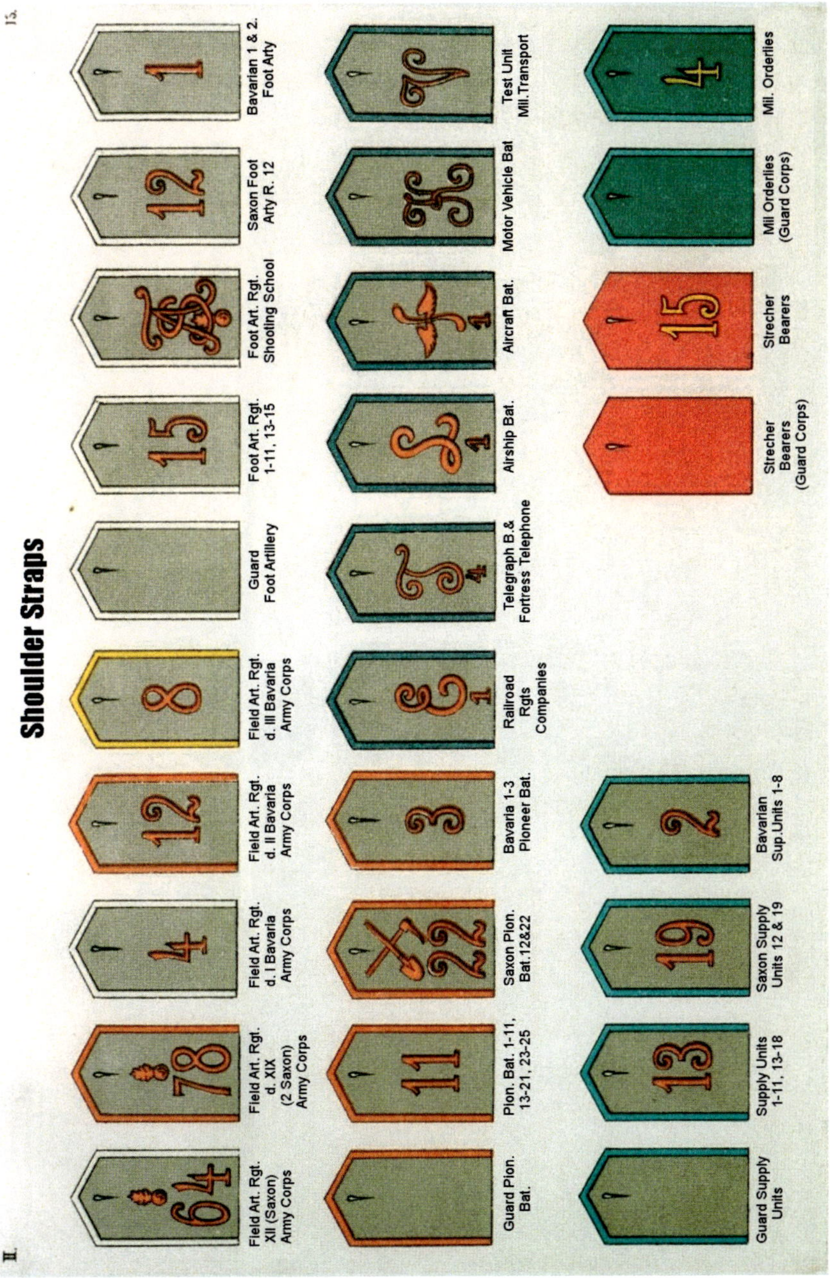
II.
Shoulder Straps
15.
Field Art. Rgt. XII (Saxon) Army Corps
Field Art. Rgt. d. XIX (2 Saxon) Army Corps
Field Art. Rgt. d. I Bavaria Army Corps
Field Art. Rgt. d. II Bavaria Army Corps
Field Art. Rgt. d. III Bavaria Army Corps
Guard Foot Artillery
Foot Art. Rgt. 1-11, 13-15
Foot Art. Rgt. Shooting School
Saxon Foot Arty R. 12
Bavarian 1 & 2. Foot Arty
Guard Pion. Bat.
Plon. Bat. 1-11, 13-21, 23-25
Saxon Pion. Bat. 12&22
Bavaria 1-3 Pioneer Bat.
Railroad Rgts Companies
Telegraph B.& Fortress Telephone
Airship Bat.
Aircraft Bat.
Motor Vehicle Bat
Test Unit Mil. Transport
Guard Supply Units
Supply Units 1-11, 13-18
Saxon Supply Units 12 & 19
Bavarian Sup.Units 1-8
Strecher Bearers (Guard Corps)
Strecher Bearers
Mil Orderlies (Guard Corps)
Mil. Orderlies

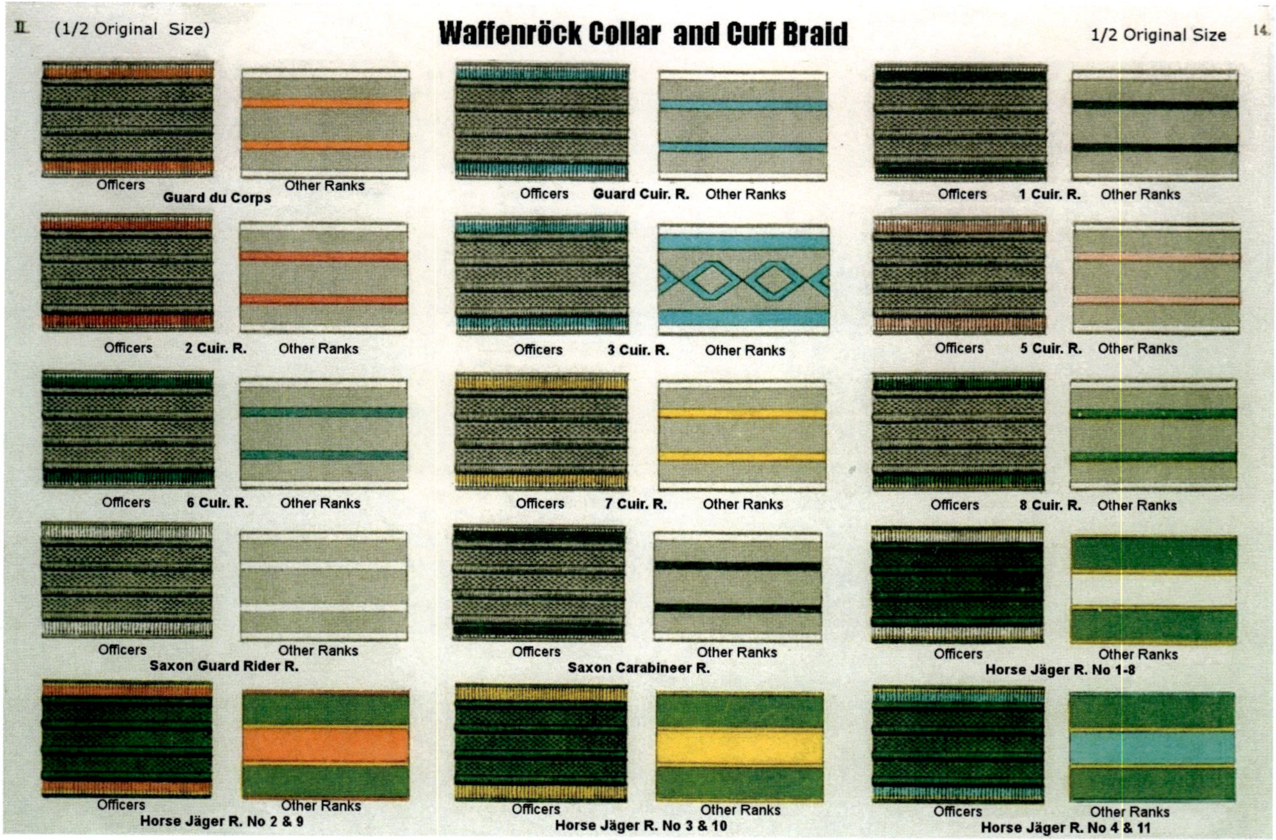
II
(1/2 Original Size)
Waffenröck Collar and Cuff Braid
1/2 Original Size
14.
Officers
Other Ranks
Guard du Corps
Officers
Guard Cuir. R.
Other Ranks
Officers
1 Cuir. R.
Other Ranks
Officers
2 Cuir. R.
Other Ranks
Officers
3 Cuir. R.
Other Ranks
Officers
5 Cuir. R.
Other Ranks
Officers
6 Cuir. R.
Other Ranks
Officers
7 Cuir. R.
Other Ranks
Officers
8 Cuir. R.
Other Ranks
Officers
Other Ranks
Saxon Guard Rider R.
Officers
Other Ranks
Saxon Carabineer R.
Officers
Other Ranks
Horse Jäger R. No 1-8
Officers
Other Ranks
Horse Jäger R. No 2 & 9
Officers
Other Ranks
Horse Jäger R. No 3 & 10
Officers
Other Ranks
Horse Jäger R. No 4 & 11

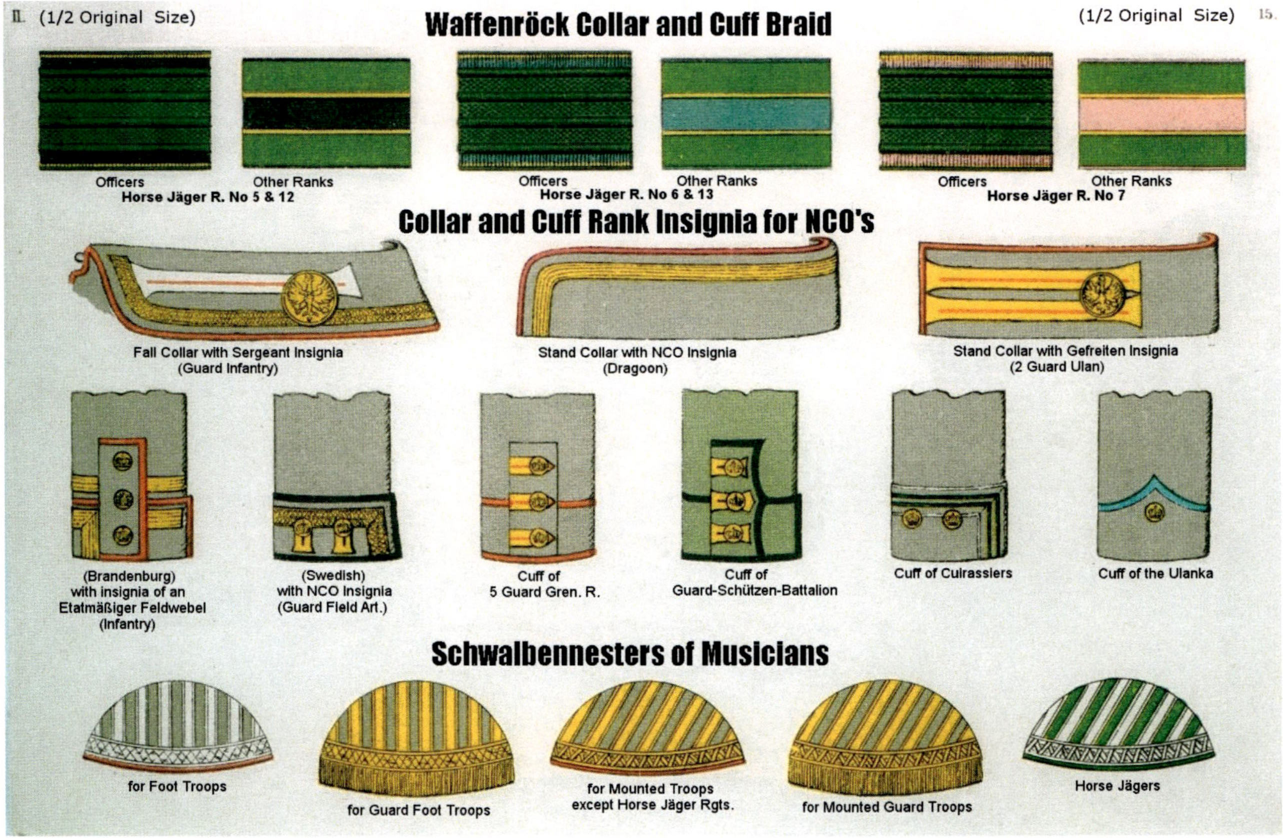
II. (1/2 Original Size)
Waffenröck Collar and Cuff Braid
(1/2 Original Size) 15.
Officers
Other Ranks
Horse Jäger R. No 5 & 12
Officers
Other Ranks
Horse Jäger R. No 6 & 13
Officers
Other Ranks
Horse Jäger R. No 7
Collar and Cuff Rank Insignia for NCO's
Fall Collar with Sergeant Insignia
(Guard Infantry)
Stand Collar with NCO Insignia
(Dragoon)
Stand Collar with Gefreiten Insignia
(2 Guard Ulan)
(Brandenburg)
with insignia of an
Etatmäßiger Feldwebel
(Infantry)
(Swedish)
with NCO Insignia
(Guard Field Art.)
Cuff of
5 Guard Gren. R.
Cuff of
Guard-Schützen-Battalion
Cuff of Cuirassiers
Cuff of the Ulanka
Schwalbennesters of Musicians
for Foot Troops
for Guard Foot Troops
for Mounted Troops
except Horse Jäger Rgts.
for Mounted Guard Troops
Horse Jägers

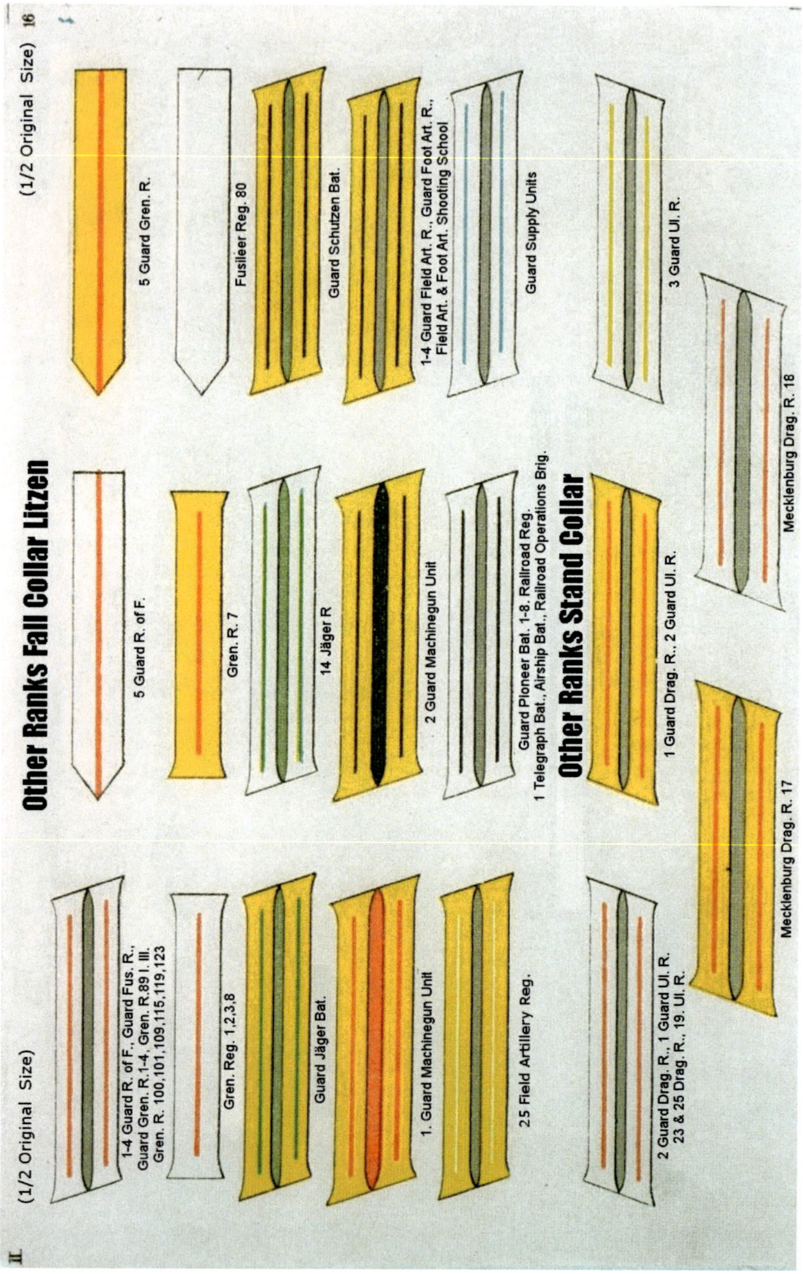
II
(1/2 Original Size)
Other Ranks Fall Collar Litzen
(1/2 Original Size)
16
1-4 Guard R. of F., Guard Fus. R., Guard Gren. R.1-4, Gren. R.89 I. III. Gren. R. 100,101,109,115,119,123
5 Guard R. of F.
5 Guard Gren. R.
Gren. Reg. 1,2,3,8
Gren. R. 7
Fusileer Reg. 80
Guard Jäger Bat.
14 Jäger R
Guard Schutzen Bat.
1. Guard Machinegun Unit
2 Guard Machinegun Unit
1-4 Guard Field Art. R., Guard Foot Art. R., Field Art. & Foot Art. Shooting School
25 Field Artillery Reg.
Guard Pioneer Bat. 1-8. Railroad Reg. 1 Telegraph Bat., Airship Bat., Railroad Operations Brig.
Guard Supply Units
Other Ranks Stand Collar
2 Guard Drag. R., 1 Guard Ul. R. 23 & 25 Drag. R., 19. Ul. R.
1 Guard Drag. R., 2 Guard Ul. R.
3 Guard Ul. R.
Mecklenburg Drag. R. 17
Mecklenburg Drag. R. 18

Appendix

Glossary of German Terms

1. **Achselstuck** – Literally "Shoulder Pieces". Officer's shoulder pieces on the field uniform for both junior and senior officers.
2. **Achselschnure** – Full aiguillettes.
3. **Attila** – Hussar's tunic.
4. **Chevaulegers** – Bavarian medium cavalry.
5. **Dienstmutz** – Service cap.
6. **Etatsmassigen Feldwebel** – Sergeant major in the infantry.
7. **Etatsmassigen Wachtmeister** – Sergeant major of cavalry and artillery.
8. **Fangschnure** – Shoulder cord for an aide de camp.
9. **Faustriemen** – Like the saber tassel but for cavalry, artillery, and train.
10. **Feldbinden** – Lace waistbelt for officers.
11. **Feld Mutzen** – Field Cap.
12. **Sergeant** – Sergeant.
13. **Feldwebelleutnant** – non-commissioned officer given a lieutenant's position without full status as a commissioned officer.
14. **Gefreite** – Private
15. **Gosch der Marine** – Imperial Navy Jack
16. **Halsbinde** – Neck Stock
17. **Halstuch** – Neck Cloth
18. **Hoboisten** – Bandsmen
19. **Hornisten** – Bugler
20. **Jäger** – Light Infantry (lit. Hunter).
21. **Jäger zu Pferd** – Mounted jägers.
22. **Koller** – Tunic worn by cuirassiers.
23. **Kolben embroidery** – Special pattern collar litzen.
24. **Krankenträger** – Stretcher Bearers.
25. **Krankenwärter** – Medical Orderlies.
26. **Landwehr** – Militia

27. **Leutnant** – Lieutenant
28. **Litewka** – Undress jacket worn with the blue uniform. The 1903 pattern for all officers was a light gray double-breasted jacket with special collar patches. There were also single breasted, fly front, undress jackets called "litewka" worn by other ranks, particularly Landsturm formations.
29. **Litzen** – Patches shaped like false buttonholes.
30. **Offiziersstellvertreter** – Officer
31. **Portepee** – Sword knot.
32. **Regimentstambour** – Regimental drum major.
33. **Reithosen** – Other ranks riding breeches.
34. **Säbel** – Sabre.
35. **Säbeltasche** – Black leather pouch bag with a black leather flap worn behind the sabre.
36. **Säbeltroddel** – Sabre tassel for infantry, foot artillery and pioneers.
37. **Schutzen** – Light infantry, similar to Jägers.
38. **Schwalbennester** – Musician's shoulder pieces.
39. **Schwere-Reiter Regiment** – Heavy Cavalry regiments used by Bavaria and Saxony.
40. **Seitengewehr** – Bayonet.
41. **Spielleute** – Drums and bugles for the infantry.
42. **Stabordenn** – Staff Orderlies.
43. **Stiefel** – Boots
44. **Steifelhosen** – High Boot Breeches.
45. **Tambouren** – Drummers
46. **Trumpeters** – Trumpeters and other musicians in the cavalry.
47. **Train** – Logistics Corps.
48. **Tuchhose** – Cloth trousers.
49. **Uhlan** – Lancer.
50. **Ulanka** – Lancer tunic.
51. **Unteroffizier** – Corporal
52. **Verkerstruppen** – Transportation Branch incl. Air.
53. **Vizefeldwebel** – Rank below etatmäßiger feldwebel and above sergeant in the infantry.
54. **Vizewachtmeister** – Rank below etatmäßiger

feldwebel and above sergeant in the cavalry or artillery

55. **Waffenrock** – Tunic.
56. **Werkmeister** – Foreman.

Made in United States
Orlando, FL
05 November 2026

90200291R10043